Renewable forms of energy

In many forms of global use

And a completely wrong green transition

But there are solutions and future possibilities are here

FSC
www.fsc.org
MIX
Papir fra
ansvarlige kilder
Paper from
responsible sources
FSC® C105338

Renewable forms of energy

Translated from Danish version September 2023

Proofreading: Jørn Nielsen

Cover Jørn Nielsen

Pictures Jørn Nielsen

Author Jørn Nielsen

Editing Paul Erik Pedersen

Layout: Paul Erik Pedersen

Publisher BoD – Books on Demand, Hellerup, Danmark

Print BoD – Books on Demand, Norderstedt, Tyskland

1st edition December 2023

ISBN: 9788743056201

Table og Contents

Preface

This book was made because, at a very young age, I began to be interested in everything alternative, and before the age of 20 I saw an older man making a windmill. It was around the time when they started on the large 2Megawatt wind turbine in Tvind, which was managed by the traveling college with founder Amdi Pedersen.

Here I met someone who could help me - and without Tvind's help, I probably wouldn't have learned any more, and later I became very interested in all forms of renewable energy.

I suddenly met a group at Thy called Nive, which means North-West Jutland Institute for Renewable Energy.

Here I met several people, including my friend Ian Jordan, which ended with us building a small windmill, after I read his translation of the "United Nation Report", which was originally done in Danish, but nobody wanted to publish engineer Johannes Juuls report on the Gedser mill because he claimed that wind energy was better than coal.

The Indians made the English version, and in the eighties, when I was 23 years old, I read Ian Jordan's translation, and it gave me a lot of knowledge.

Ian Jordan told me on 5 August 2023, when I visited him, that the report in 1961 was picked up by Preben Maegaard abroad, and Ian Jordan translated the report.

I read this report with great interest and found a formula for how to calculate a wing that is equally wide all the way out. We jointly built the first Thy mill with wooden blades, and to this day the blades still look like that but are now made of fiberglass.

I got help from Tvind's energy office by Jacob A. Chr. Bugge, who was then a leader in aerodynamics for wind energy, and who wrote the book: "The Book on Wind Turbines".

I published a 15-page booklet on how to build a windmill. It was loaned to more than 100 libraries and sold to several manufacturers, but unfortunately no one told me that to get loan money from the libraries, it must be at least 30 pages long.

My aunt typed it on a typewriter, and I made drawings for the booklet. A few years ago, I borrowed it from the library and had it copied. Today it can be found, among other things, at DTU and 2 other libraries and can still be used because, as at the time, I was 40 years ahead of my time.

I called the booklet in Danish "Forenklede beregninger til brug ved bygning af mindre vindmøller".

In these times, when everyone is talking about a green transition and CO2 neutral transition, I and other equal people believe that we are doing it all very wrong and difficult, and in this book, I want to describe how a country becomes independent from other countries and self-sufficient.

With, among other things, experience from the use of Russian natural gas and the Ukraine war, shows with all clarity that we still have not learned from the past - especially knowledge from the world wars, and the third world war, which, I believe and know, has started now.

In this book I will mention forms of renewable energy, as I have also worked with solar energy and other forms of energy, including 10 years in the Philippines.

If there are any mistakes in the book, I apologize profusely. I have endeavored to be factual, as it has been difficult to find information many years back and to remember everything in detail.

Happy reading to everyone we talk to.

Introduction

As described in the foreword, I started very early to take an interest in things that people only started talking about 40 years later. I still feel like I'm years ahead of my time.

When we started making energy systems, each house had its own system with small systems that could be self-sufficient.

Now everything must be very big and controlled by big companies, and you forget that independence from other countries, such as Russia, China and other dictators has total influence on the world.

It's a nice thought to have free trade and share everything and trust that people are not evil and will help each other. But unfortunately, the world is different.

But history has taught us that this only works for about 2 generations. After that, everything repeats itself again.

We have not learned from World War II. Just look at how China and Russia want to dominate the world and other dictators are doing the same. And the US is defaulting on its debt, widening the gap between rich and poor.

Of course, we must work together and do our best to bring the world into harmony, but at the same time we must make ourselves independent of large nations and renewable energy from outside. The so-called green transition is currently being built up like crazy.

Therefore, we all must be self-sufficient within our national borders and must change the perception that the bigger the better, is not always the solution.

I really want a world where people trade freely with each other. This can be done while being independent as a nation. In this book, I will explain how this is possible.

Figure 1: The house is self-sufficient with alternative energy with battery storage. This owner lives close to Bjerringbro, Denmark.

A little advance history from Nordic Folkecenter for Renewably Energy

I recently visited the Nordic Folkecenter for Sustainable Energy, Kammersgaardsvej 16, Sdr. Ydby, 7760 Hurup Thy, Denmark and updated my knowledge here. Below are a few pictures from the museum.

Storage of Energy

Storing energy is the main problem in the green changeover, which we are now going through, especially in Denmark. We are trying to transform our society so that everyone drives electric cars.

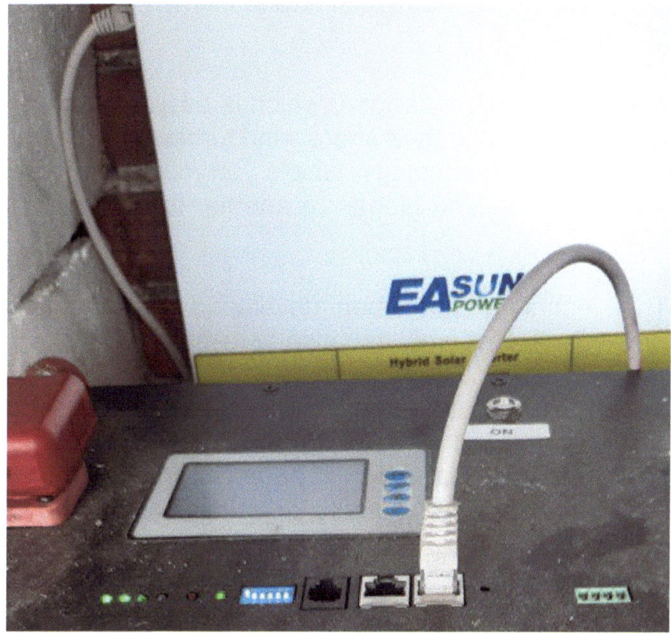

Figure 2: A Chinese battery of 48 volts is used to store the house's electricity consumption. There are alternative Danish storage batteries, but these are somewhat more expensive according to the owner.

In principle, electric cars are not wrong, but the batteries that are now used are made from raw materials extracted in the poor countries and hotspots in the world, which pollute enormously during extraction, which is not very good environmentally.

I know that the big companies and many politicians do not agree with this. But a reasonable balance can be found. We are forced to find this balance.

In time, I believe that improved technology can optimize the development of batteries, which require a different use of the earth's minerals. I sense calcium as an element.

But why not use hydrogen, which is easy to produce, especially with large offshore wind turbines and an already known technology, where you make liquid fuel with hydrogen and biogas or another form of similar gas.

This technique has been developed and is called Power to X. Methanol can also be used here.

Then existing cars, even planes and container ships can function, and it does not burden the environment. And remember that only rich people can afford an expensive electric car. We also must think about people with low incomes.

For me, it is important that we have the social aspect, so that all people, even those with low incomes, have a chance. Environmentally friendly batteries with a longer lifespan are much better for the environment than current batteries.

When used cars can run on this fuel, we solve a major societal problem, since many people on low incomes can only afford a used car. I belong to this group.

It is far too easy for politicians with their large incomes to make laws, as most of them have never tried to survive on a low income.

The same politicians believe that in sparsely populated parts of the country you can't use buses and trains, because there are almost no trains or buses.

So, in these areas of society, private cars are essential. I will now describe the problems that Denmark and several other countries have when large solar and wind energy projects are implemented on the electricity grid and here come up with solution models and innovative thinking.

Figure 3: A Chinese battery of 48 volts is used to store the house's electricity consumption. There are alternative Danish storage batteries, but these are somewhat more expensive according to the owner.

Sun and Wind

Right now, large solar cell systems and giant wind turbines are producing electricity for the grid. This is also ok, but to cover a country's needs, the facilities must be larger than the country's real needs.

This requires a very strong electricity grid with large high-voltage masts and enormously thick wires to transport the energy across the country and the excess production out to consumers in neighboring countries.

This means that if, for several reasons, no consumer can take the power from these large installations, the large offshore wind turbines must be shut down and stopped.

Many countries have previously purchased Denmark's energy made from solar cells or wind turbines, such as Norway and Germany.

But for political reasons, Germany closes the import of electricity from Denmark when their own plants produce enough.

For several years, Norway has bought cheap power from Denmark through a large cable from North Jutland to South Norway. This happened at a low price, and when we did not have enough energy from wind and sun, they opened hydropower, and Denmark then paid a high price for Norwegian electricity.

But climate change has meant that they do not have enough water in Norway, and therefore they can’t and will no longer deliver to us.

As a large country spanning kilometers from north to south, Norway has never needed to be connected by electric cables between South Norway and North Norway because each half was self-sufficient.

But now Southern Norway is in deficit with water and has problems being self-sufficient.

They have now realized that a large electricity cable must be built between these parts of the country, and this will take many years to build, as kilometers of canals must be built in the Norwegian mountain landscape.

We have politically decided in Denmark that we will direct the current towards Zealand and further down towards Poland, but the problem is that the electricity grid is not stable enough for this gigantic effect.

All this means that you must stop the large wind turbines, since no one can or will take away our surplus production.

Right now, at the time of writing, the big companies and our foolish politicians dream of making energy islands with huge wind turbines and solar panels a production that corresponds to 150 percent of Denmark's energy needs.

This is an excellent idea, but they still haven't solved storage for later use when production drops.

All this is because you have a sick idea that the bigger the better, as big capital is only interested in getting even bigger.

And unfortunately, our own companies and electricity companies are bought up by large foreign companies, and we lose control. Let's find a better balance between big and independence through local small systems.

In the same way that the big technological giants like Facebook, Apple and other big Tech giants don't want to pay taxes and don't care about poor people.

The companies are being bought up by the large foreign giants who previously dominated oil and coal, and this pushes prices up to an artificial level.

I'm not saying these companies are only bad, because we need entrepreneurs.

But if a society is to function well, it is necessary to equalize the difference between rich and poor and abolish deep poverty in the world (balance poor and rich).

Globally, the green transition can only be a success if we abolish poverty, and there is no need to have so many children in the poor countries of Africa, Asia, and similar countries.

In these countries, many children are born, who later must support their parents. But our planet can’t handle and feed such a large population, which is why a change is necessary.

When I lived in the Philippines, a Dane, Uffe, who was number 2 in UNICEF, did an experiment where they prevented children under the age of 2 from dying.

Many children were born, especially in mountain areas, as it was normal for several infants to die before the age of 2. So many children are necessary to support the parents later when they get old.

When they finished the experiment in one area, the number of children born dropped to about half.

I am also aware that this will not happen in 5 to 10 years, but we must start somewhere, and right now I am living in Denmark and therefore start by describing how we can do better here.

Later I will give suggestions and ideas for other countries and describe what I have done and experienced from the 10 years I lived in the Philippines from 1998 to 2008.

I now go back to my youth, where around 1976 and a little later I read Ian Jordan's description of the Gedser mill and its predecessor the Vester Egesborg mill.

The background is very interesting, as engineer Johannes Juul, who then worked for the Zealand electricity company SEAS in 1947, developed the Vester Egesborg mill and later the Gedser mill, which was 200 Kilowatts and delivered electricity to the grid without problems for more than 10 years.

Johannes Juul claimed in 1957, the year I was born, that wind energy was better and cheaper than coal power, which was used at the time. He was a wise man and was ahead of our development today.

He also claimed that wind energy was better and cheaper than coal power, which was then the most important energy source in Denmark.

As you can probably guess, it did not resonate, and people thought that the man was a bit of an idiot to say that we should move away from coal power.

He developed the Gedser profile and made wind tunnel experiments, and you can read that he put sticks into the snow and saw how the snow folded in the wind around these wooden sticks.

He was also the one who solved the problem of how to prevent a wind turbine from running wild and goes to overdrive in a very simple way by letting the outermost 10 percent of the blade rotate across the wind.

This was solved by means of a spring inside the wing, which due to the centrifugal force via an oblique rotation in a metal tube, where the tip of the wing turns, and can't turn back until the revolutions are almost at zero.

You know the principle from airplanes, which in technical language stall, which lose buoyancy and can crash.

A few months ago, a very large wind turbine had run wild and posed a great danger to the surroundings. Attempts were made to stop it by throwing a strong rope and net over the wing in the hope of slowing it down. Fortunately, nothing happened, and was later brought under control.

Today, people turn the wings at the root, but they do not think that if more accidents happen with the automatic system, the forces are too great.

Unfortunately, I can't remember the year, but in Sparkær, close to Viborg in Denmark, there was a very skilled designer and manufacturer of fiberglass wings up to 5 meters in radius.

His name was Erik Grove, and he was also inspired by the Gedser Mill and the Risager Mill. During this period there were many talented young people who greatly shaped our future.

A 5-meter blade was then used for the turbine in the eighties and had an output of approximately 22 kilowatts. The wings were produced in Sparkær by Erik Grove.

These turbines were all connected to the grid and ran at a constant speed, and here an accident happened. A wing ran wild in stormy weather and came up to high revolutions.

One wing flew very far away, and as far as I remember about 2 kilometers, where the wing landed with great force in a garden behind a house and plowed vertically into the ground.

The incident could have destroyed several houses, and had the wing hit a densely populated town, it could have cost human lives and great destruction, but the horror was spared then.

The blade manufacturer talked about this at an energy meeting where I was an audience member. He and others had subsequently made calculations on how much power and speed (rpm) the wing was at.

It was calculated at the time that in just 6 to 7 seconds the effect increased from the approx. 22 Kilowatts to over 700 Kilowatts with extreme speed at the tip of the wing.

Shortly after the incident, the manufacturer's blades were made, as Johannes Juul described at the Gedser mill, and the turbines could not run wild.

This has been forgotten with the huge wind turbines that are being made now, and here I must point out that safety is more important than optimizing performance to the maximum.

Johannes Juul (1887-1969) tried to make a report in Denmark about wind power, but the authorities and the energy companies did not want to publish it, because the man claimed that we should not use coal power.

But then India approached him, because they wanted wind power in India, and it ended up being published in English as a United Nation Report for them.

My friend, Ian Jordan in Thy, where he still lives, later found the report abroad and translated it into Danish. As previously described, this happened together with the pioneer Preben Maegaard.

I read the report and in 1979 and I published the booklet: in Danish "Forenklede beregninger til brug ved bygning af mindre vindmøller".

Soon after, my formulas were used, and when THY Mill was built after my book. The wings were with the Gedser profile, as far as I remember. The first wings that we produced were made of wood. I was part of the project.

We made hot water for a hot water tank with electric cartridges and three-phase power with an asynchronous motor, which was a standard three-phase motor.

The first turbines we made were not connected to the grid, but used capacitors of 400 volts AC, which created the magnetic field.

A very small magnetism in the rotor meant that at about the speed for which the motor was made, it went into resonance, whereby a very high voltage came to the terminals, where you normally connect three-phase current of 380 volts used as a motor.

I developed a simple controller which switched the heating elements into 3 steps or 6 steps depending on the voltage that the motor gave off. If there is no load example heat elements - the three-phase voltage on the motor, which is our generator, will rise to a very high value and destroy something.

But the control allowed us to adjust the load so that the voltage was neither too high nor too low to provide enough heat in the storage tank.

Racing could not happen as we could control both volts and revolutions with extra heaters, and we could also set the revolutions down on the wings.

This was done by a relay connecting - a set of extra capacitors.

This changed the resonance point and the revolutions at which the engine produces energy fell.

This is a short explanation, but we now had a cheap off-grid turbine which, together with a home-made solar collector made with a black-painted radiator, could produce hot water for domestic water and heat for heating a house.

We made systems where we used a 1200-to-1800-liter oil tank as storage with a heat exchanger at the top of the tank for flow temperature for radiators or better underfloor heating and a welded hot water tank at the top of this oil tank for service water (example for shower).

Materials were used which were normal in the plumbing and oil industry.

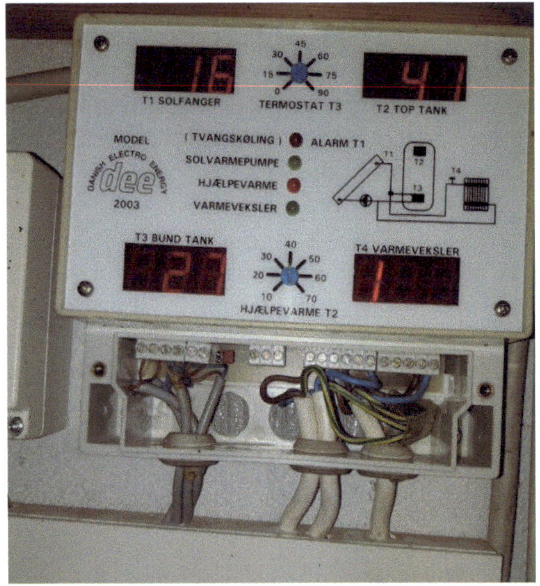

Figure 4: Own manufactured control for solar heating, developed for self-build projects in the 90s.

I developed several controls with digital readout of sensors which could read the temperature of the solar collector and the temperature of the tank both at the bottom and at the top. Only one controller had my own name on it, the rest were manufacturers with their companies' names on it.

In the 90s, I had 20 percent of the Danish market with controls for solar heating. We struggled with the subsidy for the industry coming and going, which meant that sales dropped a lot during the periods when they removed the subsidy for solar heating. It cost the lives of some fabricators and installers.

I will return later with the knowledge that the Gedser mill report gave us, which started a lot of this, and I will come up with solutions to our global green transition problems to the best of my ability.

Figure 5: Smaller private Wind turbine with very few kilowatts.

Today there are major problems with obtaining permission to set up a small wind turbine, and this means that many who have a solar heating system cannot become completely self-sufficient with heat.

But permission is given to very large wind turbines that run online in the same areas. This bothers many people who live in the countryside.

If we mean anything by what politicians claim about being green and saving the planet, in my opinion, you should also build many small systems.

The pictures below show small plants for independent and self-sufficient small dwellings in the countryside, and the bottom picture with the solar cell plant, which can cover large urban communities.

Here the difference between self-sufficient facilities and collective large-scale operations is shown. See pictures below.

Figure 8: A smaller private wind turbine is shown here, reminiscent of the Thy windmill for a smaller house.

Figure 7: Typical solar system for a private house.

Figure 6: Gigantic solar system that can cover several major cities.

Small and large Units

I will now come up with ideas for how to store energy with both small and large plants. Now let's look at the storage of electricity produced from large wind farms and large solar parks.

Let's start with small systems for a small house, it could be with solar cells on the roof and possibly also a small wind turbine, which produces 12 or 24 volts for a battery.

I would suggest that you replace all lighting with LED lighting that runs on 12 or 24 volts, as well as replacing refrigerators and freezers with a model that can also work on 12 or 24 volts.

Today there are battery stores that can supply a house with these types of energy, and with the combination of sun and wind and a large battery store, it will be able to function all year round.

Some will then say, what if the weather for a period means that there is not enough energy, what do we do?

The solution could be to charge the batteries via a charger from an emergency generator, or via power grids to have a charger which comes into operation here.

About space heating, as previously described, you can have a large storage tank which can get its energy from several forms of energy. It can be a solar collector or wind turbine or a wood stove or pellet stove.

If you design your house or change it to full or partial underfloor heating, a much lower supply temperature is required for the water that is to heat the room.

Since it is easier to produce a low temperature from a solar collector, you get a greater performance, and most of us find it very pleasant to have heat on the floor in the bathroom, and it could also be in other rooms.

The idea here is to utilize several forms of energy and use as little supplemental heat as possible and at the same time live a comfortable life. With such a system, an electricity grid is not so strongly required in several areas of the country.

This may not be the whole solution, but there are many solutions, and they will be described later, and I hope that others will start thinking in new ways so that we have a sensible green transition.

Now I want to come up with proposals for how we store the energy from large facilities that are connected to the grid from large offshore wind turbines and large solar parks.

As described earlier, a powerful power grid and consumers are required for such an energy system. If more and more people switch to electric heating or heat pumps, problems arise at certain times of the day or seasons.

Water is a good energy store, and so are large tanks of compressed air. Unfortunately, batteries have not yet been invented that can solve the consumption of an entire country as previously described, but I think it will come.

If we make some lakes in a high place, which could for example be on the Jutland ridge and a little lower in other lakes, we will be able to solve some of the problem. This applies to Denmark, but everything can be adjusted in other countries.

In these artificial lakes there can also be beautiful nature in and around. When the large wind turbines produce a lot of power in windy weather or solar parks, we pump the water from the lower lake up to the higher lake.

Our well-known company Grundfos or other companies make pumps with very high efficiency - typically 90 percent or better. So, there are good pumps on the market.

When there is not enough electricity, we let the water flow down to the lower lake, but not with too much pressure.

Better with many small and slow-rotating generators connected in parallel, which then supply energy when there is not enough electrical production.

We must also utilize our rainwater sewers. Worse, there is a tendency not to make the pipes large enough for future use. Here, generators in the rainwater pipes could also produce electricity.

And when you build new buildings, you always forget to make it so that the rainwater on the roof cannot run over the edge.

Make the roof so that there are large gutters and edges on the sides, so that all the rainwater ends up in the drainage pipes. But better with larger drainpipes.

At the bottom of the building, slowly rotating generators can produce electricity.

Perhaps these generators must store the current on batteries used for LED lights and refrigerators and freezers, which could be 12–24-volt systems. And 48-volt systems can be used for larger installations.

One could also utilize these large volumes of water and lakes as an energy source for a heat pump system for district heating.

To get the greatest possible result from this method and other methods, it is best to divide the country into areas that are not too large and make them self-sufficient.

These many areas can then be supplied via the electricity grid, so that they can borrow energy from each other and deliver to each other. Many small communities and rural areas then get better coverage.

Now for the next storage solution, which is Compressed Air. But first in the next chapter a little history.

Figure 9 Here we have a section of a larger plant that supplies hot water to a district heating plant, in connection with a biogas plant.

Compressed Air

When I was in my twenties, in Jutland, Denmark they wanted to store nuclear waste in the salt horsts in metal containers, which were to be deposited in the underground.

Although these metal containers were also coated with a cement layer, it was not long-lasting. These containers were to be deposited in the Jutland salt horsts.

Over many years, they would be eaten away by the salt, and the radioactive material would seep into the underground, where all of Jutland's groundwater is, and here destroys our drinking water, as the decomposition time is many thousands of years.

I was part of the citizens' group against nuclear waste, and we manage to prevent this. No one has yet disposed of nuclear waste properly anywhere in the world.

I know a lot of people think I'm wrong on this one. Let's hope future generations fix this. And future generations will have to find a solution to our stupidities.

The publisher of this book has told me that he thinks we should send our nuclear waste out to the sun on a rocket and burn it up there. Good idea, but not realistic and probably very expensive.

In Finland, they have expanded nuclear power, and here they will deposit the highly radioactive waste underground in the mountains. Let's hope that nothing goes wrong and that future generations find a workable solution.

But now my idea is to have a store of compressed air in the ground and thus store energy from our large wind turbines and solar cells.

If you build underground - especially in areas where there are already underground holes or make holes in the ground - large stainless steel pressure tanks with concrete around them for stability, compressed air can be stored.

The surplus flow then pumps compressed air into these tanks via compressors, and when electricity is needed again, the compressed air is opened.

This compressed air drives one or more generators, which can send electricity out into the grid locally or perhaps to larger areas of the grid.

I want these compressed air tanks underground for safety reasons, and at the same time you don't destroy the landscape.

I am convinced that this will work well and to the purpose. Together with lakes as energy storage, as described before, it solves big problems.

Figure 10: Energy lakes with a difference in height, which can be used for an energy store. Here as Hydro power.

Hydrogen

As described earlier, hydrogen will also be perfect, since the technique has been invented, and in countries such as Germany, many buses run on hydrogen.

Water is split into hydrogen and oxygen and via a chemical process - where you use gas, which could be biogas or another gas - a liquid fuel is produced. Alternatively, you could use vegetable oil and extract fuel.

This fuel can be used in countless places. You can burn it off and run some generators, or you can use it for transport.

It creates almost no pollution and provides a good CO2 balance and replaces coal, oil, and uranium, which are dangerous for the environment in the long term.

Environmental activists are usually good and reasonable people, but I don't understand their opposition to this.

And then it can also be used for district heating. If you burn it off to make heat for district heating, it is easy to clean the smoke by passing the smoke through water and then cleaning the water.

I therefore do not see anything particularly polluting in this process. And I am convinced that it will be good in poor countries, and it will eventually remove coal from our global use.

The method, which has already been invented to connect hydrogen to gas and make liquid fuel, will soon be used on Maersk tankers, and solve a major pollution problem.

But, as mentioned in my introduction, it can be used in all areas of transport such as passenger cars, planes, buses, and trucks.

Right now, it is more expensive than current forms of energy, but with the removal of taxes or subsidies over many years, it will make the green transition better and faster.

Right now, Tesla is the leader in electric cars and innovation, but it is expensive and the raw materials, which are rare metals, are polluting to extract and recycle.

After all, there have been electric cars that caught fire in them, and special equipment had to be used to put out the fire and there was no way to cut a passenger free in the event of a serious traffic accident resulting in fire.

In addition, it is easier and faster to fill cars with liquid fuel than to stand at charging stations for a long time. Let's use more combination techniques.

There is a big difference between whether you spend 5 minutes filling your car with liquid fuel or must wait 15-20 minutes. on a quick electric charge, or what is usually 2-3 hours.

And yes, in time we will develop good batteries, and I intuitively sense that calcium will be used in a few years.

Now that we are talking about the transport industry, there are also trains, buses, and large trucks.

Thomas Edison, who helped develop the direct current motor, already dreamed at that time that trains would run on batteries.

But then a solution was invented where crude oil could be turned into diesel and petrol, and his idea was not realized.

Figure 11: Already used hydrogen technology.

Train Operation

Today, when I see the light railways in Denmark and the stupid high-voltage power lines, which are dangerous to life, supplying the trains with electricity, I feel bad. Are we living in 2023 or what?

There are often technical problems, a bit of frost and like that, and it doesn't work. Our whole way of doing train operation is wrong. Not just wrong, but very wrong, and new thinking is needed.

We have problems with track changes and signal lights that fail, and all too often you must put buses in.

We should never here in Denmark have let the Italians do this. In other countries such as Japan, it works much better.

But I see solutions, but it requires new thinking - also about security. Right now, our trains run on rails, and when an oncoming train comes, you either must wait or cancel, because idiotically there is only one track on many lines.

The solution is that there are 2 sets of rails track with a suitable distance, where the right train track runs to the east and the left train track to the west.

If the distance between the 2 tracks is adequate, no signal lights or frozen lane changes are required.

In this way, both trains can pass each other without problems, and of course there must be no transitions over the rails, so that livestock, cars, and other things can cause accidents. This could be solved by building tunnels or transitions.

Whether the trains must run on rails is perhaps also wrong, but I wonder if someone will solve it soon. A technique has been developed where magnetic fields under the train create propulsion so that the train levitates.

The question is whether the trains will run on electricity or hydrogen or perhaps both.

I imagine that the train has, as now, several carriages connected, where the passengers sit and enjoy the ride, and there is an additional train carriage connected, where the train driver and other staff are located.

This extra train car is filled with good quality batteries, and here at first it could be marine batteries that golf carts use, but later my calcium battery.

This train car of batteries powers the train, and when you reach the terminus, you remove the batteries and install new charged batteries in a few minutes or change the battery train carriage.

You can also imagine that the train runs on hydrogen some of the time and charges the batteries, or there are solar cells on the roof which also charge the batteries during the daytime, or a combination.

I imagine that one energy source sets the train in motion, and when a certain speed is reached, the next energy source steps in.

Then we won't have annoying wires like now, and the next phase could be that we no longer run on rails, but let new people work on this.

As far as I know, trains in Japan runs at high speeds, but as with everything, safety must be a high priority.

I would like to be able to say exactly how we solve everything, but this very thing, I am sure, will work, and when fresh young forces work on it, we will get a sensible green transition for the benefit of all the people of the globe.

Figure 12: Japan's high-speed train with over 600 km/h.

Figure 13: Another Japanese high-speed train.

Wave Power

The next thing I want to talk about is wave power and later water pumping, which I have been involved in when I lived in the Philippines. Finally, I will conclude with thoughts about the future energy source many years into the future.

I imagine we use the tides to make our electricity. In England, they have tried to make a system which makes use of the fact that the tide raises a plate up and down.

A large plate is anchored on the bottom in some hinges. The plate is always under water, and via a large exchange it pulls a generator.

Here, the tide rises over 12 hours up to 6 meters when the pull of the moon causes the water to rise, and for the next 12 hours the water goes down, and this cycle is stable.

Similar methods can be used elsewhere and best where there are no violent waves. In the sea not far from the coasts, you can use a system with wave power, where plates tilt up and down, and this has been tested around Hanstholm in Denmark, and it worked well.

I also imagine that we anchor a system on the sea floor or build some kind of island above the button of the sea. Here, a generator is mounted, which is driven by propellers higher up towards the attack, which gets its energy from the movements of the Gulf Stream.

Via a cable on the button on the sea, the energy is sent to land, or perhaps a ship comes at regular intervals and brings hydrogen to the mainland.

Many will probably say that now I'm struggling, but we must think new, and water is one of the biggest energy forces we have.

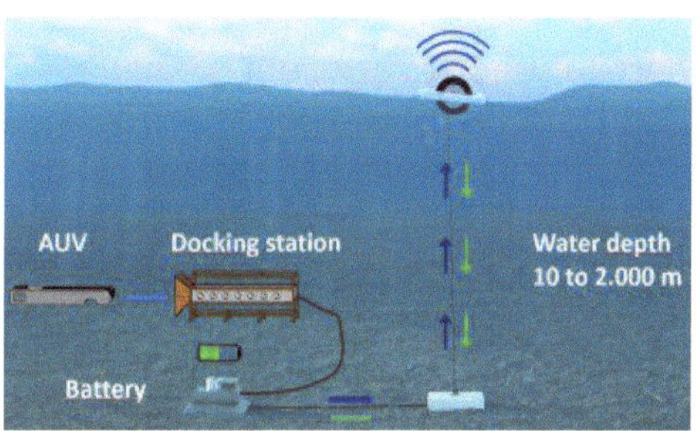

Figure 12: Utilization of ocean currents.

If we can make it so that it is not destroyed by storms and tsunamis, it is so amazing and stable that it can be used almost globally.

Figure 13: Utilization of tilting floats in the water surface.

Hydropower and Plants

In areas with unsuitable agricultural land, planting can also be used as a source of energy.

Along rivers and low-lying areas with water, you can plant willows and poplar trees and other fast-growing plants, which both hold the soil and prevent the leaching of nitrogen.

These plants are then cut back every year. I suggest that you make biogas from them or burn them off in suitable ovens, where the smoke is washed and cleaned.

This reduces CO2 in the air. Plants are part of our CO2 solution, since they absorb CO2 from the air and store it in the wood. I see that the method is close to CO2 neutral and good for the environment and nature.

In this way, it is possible to make electricity, thereby providing energy for district heating projects.

The more solutions that are used in combination - depending on a country's infrastructure - the better it will be.

It makes it easier to abolish poverty and become independent of other nations and is especially good in poor countries where it can be exploited.

No one says that the same methods must be used in all countries, but they must be adapted to the country's biology, landscape, and climate as well as the country's culture.

Right now, we know that large farms require very large areas for either grazing animals or crops for the animals.

All kinds of plants absorb CO2 from the air and store it in the plant. Many plants are good for the planet, and if we can stop the clearing of the rainforests, mother earth will be better off.

But if we are to produce food, such as vegetables and all kinds of edible plants, it usually requires large areas.

That is why we see very large areas in modern large-scale agriculture. But in small areas and local communities there is often no room for it.

And if we are to have plant production for food in cities, new thinking is required. It has already started in a few places.

Greenhouses are made, where plants grow on shelves and usually without soil, and nutrients are added, and water is recycled.

As we know from open and large areas of land, you are very dependent on it is raining at the right time of the year, and it must be dry when harvesting.

But unfortunately, the weather is not as you would like, and the harvest often fails and causes hunger in poor countries and generally with poor economy as a result.

But if it is in greenhouses, as described, the time of year means almost nothing, and you can grow and harvest all year round.

But how do the plants get the light they require, since it is usually the sunlight that ripens the plant?

The solution has been invented and used. LED light is used, which has the color that provides the best growing conditions for the plants.

And as an extra plus, you don't have to think about day and night, because the light can be on day and night.

The energy for this artificial light can come from wind or solar energy and is stored in batteries.

The technique of greenhouses also means that they can be placed inside our cities, and with a little sensible planning of roads and houses, it can be integrated into almost all cities.

It also provides stable food if war, natural disasters, or similar destructions.

If we now create such greenhouses at all our schools, it will be able to be used in the education of children, so that we will have a generation that learns self-sufficiency, thinks in new ways, and will become the next inventors.

In the poor countries, where water is deficient, you can use water from the sea, from which you remove the salt via solar energy and use the water in the greenhouses.

This will eliminate much poverty and save on the expensive fresh water that otherwise evaporates. This process will help improve and restore the ecosystem.

So, think again. In these areas, canals can lead water to our greenhouses from areas where there is more water. So, desalination of water from the sea is necessary in several places.

Which method should be used to make fresh water from seawater, I leave to colleagues and young people who want to think new things.

It will initially be expensive financially, but we also cannot live with large areas of the world not getting rain and thus creating hunger, death, and refugee flows.

If world leaders mean what they say about saving the planet, they will step in and pay for an expensive project.

The task will take many years to solve, but we have no choice as it must be done.

Basically, there is no lack of money globally, but unfortunately only an incorrect distribution. The USA is an example of this since the rich don't want to pay taxes.

Figure 14: Wetland area with plants that absorb CO2 and are used for energy production and prevent nitrogen discharge.

Earth heating

Now I want to describe geothermal energy, which is a fantastic source of energy. Geothermal heat is several things and is already used.

When you use a heat pump to produce heat to heat a house, the heat pump must steal energy from somewhere.

You can use the air and cool the air down to a lower temperature, and the compressor in the heat pump then raises this low temperature, which is used to heat water.

The hot water is used for space heating and water for baths and so on, which is typically 40 to 50 degrees.

The disadvantage of using air as an energy source is that in winter the temperature outside is cold. This results in a lower degree of efficiency.

But if, on the other hand, you take the energy from the ground - typically 2 meters down - frost does not occur, and this gives a good degree of efficiency.

This is typically done with a liquid in hoses buried in the garden, a liquid which cannot freeze, and excess energy is transferred via a heat exchanger to a container inside the house.

I have helped that we used surplus heat from a solar collector, and when it produced too much energy, it was led out into the hoses in the ground via a heat exchanger.

With this, the soil was used as storage from summer to winter, and with just a few degrees of elevation of the soil's temperature, the energy/heat in the soil tubes prevented the soil from being exhausted in the winter.

I think that if you go down deeper - depending on where in the world and which types of soil are found underground - you will be able to store large amounts of energy for later use.

In these soil hoses, it has been learned that with clay soil, a hard crust forms around the hoses in the soil. This hard crust prevents the heat from the surrounding soil from being transferred to the soil hoses.

But when, for example, excess heat from solar collectors is pumped out into the hoses, this crust dissolves again.

If you want to build large heat pump systems for urban district heating, you now use air, but as described, not the best solution. A combination is better.

Here, a solution with earth hoses could be used, but why not use the water depth of lakes and take the energy from here?

If the depth is great enough, frost never occurs, and with large areas there are enormous amounts of energy. But even better would be to use seawater for these large heat pumps.

The hoses as pipes could be buried in the bottom of the sea, where we are now approaching the interior of the earth as energy.

As previously mentioned, where electricity is made from the water in our rainwater sewers, you could also have a system that uses some of the rainwater's energy for large heat pump systems.

Yes, I know, it requires us to build differently, and completely rethink our housing construction and road structures. But the future makes it possible, so let's start now.

If we take countries like Iceland with volcanoes and hot lava in the subsoil, the subsoil's heat is already utilized here.

Near where I live in an area known as Kvols in Denmark, they tried to drill underground for geothermal energy.

Unfortunately, the project failed, but as always stupid politicians stopped this before time. But they soon start up again in new places.

The idea is brilliant and can provide stable energy far into the future, and it will never run out.

When drilling approximately 3 km. down into the underground, you come to an area where the earth's internal energy is. We live on a planet where the crust on which we live is very thin.

Deep down, the earth's interior is liquid and glowing, reminiscent like the lava that comes up from a volcanic eruption.

And those who drilled for oil before, can now use them to get down where the temperature is constantly at about 70 degrees or higher.

And the advantage is that no pumps are required, as the hot energy from the depths is pushed up to the surface by itself.

Here, the energy is deposited in a heat exchanger, which supplies the district heating network, and enough energy can be supplied for even large cities.

The cooled energy from the underground is led down into another hole, so that it can be heated again from the earth's interior.

This source of energy is almost impossible to exhaust and extremely stable. The only disadvantage is that you must have a very large network of thick, highly insulated district heating pipes, which can supply cities and dwellings far away.

There are many areas in the world that can use the method, especially if we humans collaborate globally.

I also imagine that you let the cooled liquid, which is sent back to earth, draw generators, and thereby get a stable electricity supply around the clock.

But again, there are many solutions and combinations, and we must therefore start with new thinking.

We need to invent a simple method without rotating generators that make heat energy for electricity.

So young girls and boys put on your work clothes and think new, because it is you who must invent new methods.

I imagine that all forms of radiation and wavelengths can be converted into electricity and other forms of energy.

Right now, they are working on making fusion energy, which corresponds to what happens in the interior of the sun.

Unfortunately, a very high temperature of millions of degrees is required to get the process started, and nuclear power is now used for this.

If the process is successful, you have an energy source that does not produce harmful radioactive radiation.

The first attempts did not lead to a meltdown of our globe, as many feared.

You know the problem from the weapons industry, which has developed a hydrogen bomb, which kills all living things, but does not afterwards make the area uninhabitable for a long time, as with an atomic bomb.

Do not misunderstand me. I don't like either atomic bombs or other weapons of mass destruction, but just trying to explain that the technique is borrowed from here.

Unfortunately, it is the weapons industry's products that we later use in civilian life. Solar cells started out being used for satellites that hovered above the earth and were developed by the military.

Fortunately, it has now been developed for the use of all people and let's hope that something good will come of it. See my prediction at the end of the book.

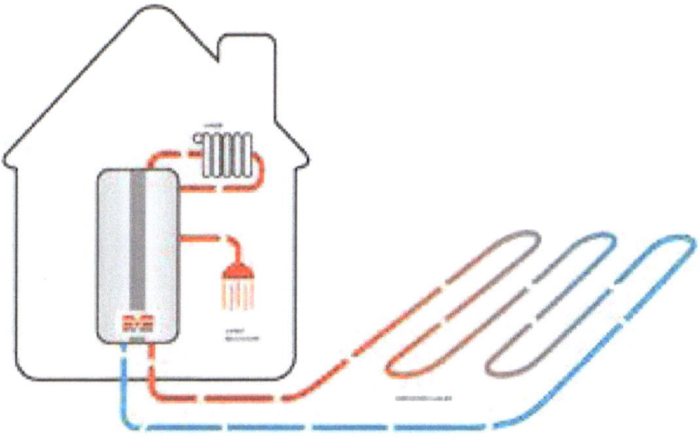

Figure 15: A typical geothermal system.

Figure 16: This is how a geothermal heating system works.

Pumps

I will now talk about water pumping and my experience from the 10 years I lived in Bauang La Union in the Philippines, 270 km north of Manila.

As everywhere else in the world, pumps made for 220 volts or three-phase electricity were used.

In the US it is 110 volts and in Japan 100 volts and Europe 220 to 240 volts AC.

These pumps worked fine when using mains power. Quite a few manufacture them.

The poor and many others did not have such a pump, instead they installed a hand pump and filled buckets with water from a well.

The electricity supply was very unstable. It was normal for the electricity to go out for a few hours every day or even whole days.

So, a solar panel was a good solution to install as well as a good quality car battery. Then you could get light when the mains failed.

But if you need to use lamps used for normal use, 220 volts an inverter are required.

These were also made locally, and I was involved in a project making inverters for buses.

Back then, it was difficult to get a television or screen mounted in a bus, which typically runs on 24 volts.

When taking the bus to Manila or another city some distance away, in almost all buses there were movies shown on a TV screen with a connected video player.

So, a inverter had to be used. After a long time, I managed to find a supplier who could supply El iron for transformers.

As with everything else, it was Chinese suppliers that supplied these things. Typically, from companies that correspond to a timber trade that also sold electrical parts for houses.

In English it is called a Lumber. When building a house, this Lumber can supply everything you need to build a house.

They have everything from cement to iron and electrical parts and water pumps.

Where we lived there were several of them and almost all were of Chinese descent.

These were typically descendants after the Second World War, and our Lumber's father started the business shortly after the Americans liberated them from the Japanese.

The son then took over the business and his children were now part of it.

And then they were fluent in the local language, perfect in English and spoke Chinese.

I was asked if I wanted to learn Chinese because then they could help with that.

I asked them sometimes why they never married a Filipino, but the Chinese demanded that it must be a real Chinese.

When I later found an importer in Manila who bought transformer iron in China, the importer told me that he had some problems with the supplier, as the Chinese supplier felt that there were not enough Chinese genes left in the family.

Just look now at 2023, when China refuses to recognize Taiwan and Putin's ideology.

We made quite a few inverters for buses but had to see that others took over and stole our products. It was a tough but also educational process. And other products were invented.

But if you were to pump water via a inverter from 12 volts to 220 volts and a standard pump, a inverter of 2000 watts was necessary.

Should a refrigerator work, the problem was the same. The price of a solar system would be huge, and solar cells were not free.

I started with Siemens solar panels, as they were the best in the world at the time, but they had to be imported, and soon after BP Solar and Shell Solar came on the market.

When these leading oil companies entered the solar market, in headquarters in Manila, there was one importer, but they demanded that you buy a package solution or nothing.

When I later tried to buy solar cells from Siemens via Singapore, the minimum requirement was too high, as far as I remember, the requirement was a minimum of 20,000 cells.

I had almost no capital, so it was impossible. But more about that a little later when something worked out.

In the USA there was a factory called Flojet that made bilge pumps for marine boats. One of my later customers and suppliers, Bruce from Canada, made a pump from a 12-volt Flojet pump that could lift water up to 50 meters and more.

The pump was waterproof, and in Danish it is a submersible pump, which is lowered into the water with a waterproof line connected to the pump housing.

But first I managed to import the Flojet pump directly from the factory in a few pieces directly from the USA.

At our house, a water tank of 600 liters was mounted on a welded stand approximately 5 meters high.

The tank is filled with water, and from the bottom of the tank a pipe goes into the house at approximately ceiling height. This gives a weak water pressure, which means that water comes out to the kitchen sink and to the toilet and at the same time makes it possible to take a shower.

At the top of the tank there is a float switch which switches the pump on when the water level is too low and switches off when it is almost full.

A 50-watt Siemens solar panel was then mounted on the water tower facing south.

This solar panel charged a normal car battery used in standard cars typically 60 to 100 ampere-hours.

The Flojet pump was mounted above ground and a ball valve sat at the bottom of the well preventing the water from running back.

The pipe was then filled with water, and when the pump was running, the ball valve opened, and the pump sucked up water.

With this technique, we could pump the water from a depth of approximately 8 meters and lift it approximately 5 meters up to the top of the water tank.

As far as I remember, the pump was 60 watts, and with the solar panel of 50 watts we could pump 1200 liters of water a day - even in the rainy season with only a little sun.

Several plants were constructed according to this principle, but then it happened that Flojet changed its dealer policy and made it impossible.

Then I bought the pump via Bruce in Canada, and sold solar regulators to him, which were partly made from local materials used for windows and electronic components bought in China and Germany - also from Denmark.

Unfortunately, the amount he bought was very small, but he was a great man.

I also supplied solar regulators to a company in the US that manufactured a solar-powered cart that warns with a flashing arrow where there are roadblocks.

I and 3 employees could live on this for approximately 3 years. The electronics were lacquered in several layers of lacquer because the requirement was that rain must not affect anything.

Unfortunately, an employee left shavings from a drill cleaning the holes that were drilled in the metal box where the circuit board was mounted.

Even though it was varnished and meant nothing to the function, it went wrong.

An American customer opened the box and saw these chips. I offered to supply new ones free of charge, but an American supplier offered a 5-year warranty and unfortunately, they ended up not buying any more.

Bruce came into the picture after this, but unfortunately, he was too small a customer.

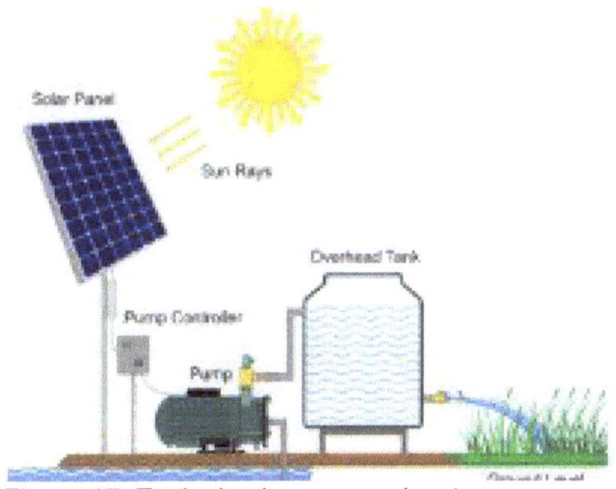

Figure 17: Typical solar-powered water pump.

LED

In 2001, the white LED came on the market, and I got in touch with a company in Hong Kong called Lucky Light.

I was sent samples, and everything was sent by DHL as the postal system was impossible. A letter from Manila to us took 3 weeks, even though there was a bus several times a day from there. The journey by bus took approximately 10 hours.

But when using DHL or Fedex - a letter or package from Hong Kong via Manila arrived at our address in 2 days.

Even though there was a meter of water on the roads during the rainy season, DHL always arrived the next day.

After several attempts, we got a white LED that was not at a light angle of 15 degrees, which is used for flashlights.

It was at 160 to 170 degrees, and when you mounted it in the ceiling, it shone uniformly and pleasantly throughout the room. This made it possible to fabricate a circuit board with some of these LEDs and mount the board in the ceiling.

We hiked around mountainous areas where there was no electricity. I clearly remember that we first drove by car for many hours north and arrived at a small village.

After that we hiked for 7 hours over several mountain peaks and arrived totally exhausted at a small village with a few houses.

We brought a solar panel and our LED light and some tools in a backpack as well as a car battery. When we arrived, they asked if we wanted to go over to the next small town and see how they lived.

But none of us could take it anymore. We spent the night. Chickens roamed freely and we were asked which one we wanted to eat and then it was slaughtered in front of us.

After a few hours of sleep, some breakfast, and a chat, we installed the solar panel. We also had a cell phone charged with the panel.

We had the light installed in the ceiling, and now the children could sit and do their homework, so the family no longer had to use kerosene lamps or candles.

The sun rises at 6 in the morning, and at 6 in the evening it is completely dark.

Several plants were later made according to the same template. And funnily enough, you could buy Coca Cola up there, just like out in the jungle.

We asked where and how they got Coke and other things brought there. The answer was shocking. We hike down to the village where our car was parked and then we carry it up.

They did this in 2 hours a couple of times a day without getting tired where we were smashed after just seven hours walking over several mountain peaks.

And there were children who walked just as far to go to school.

These led diodes are made in different light scattering angles from 15° to 170°.

If you want to create perfect room lighting, choose a large angle, and they are available in different colors, from warm white to bright white light, which is close to bluish.

Furthermore, they are available in different colors which can be used for lighting plants in greenhouses.

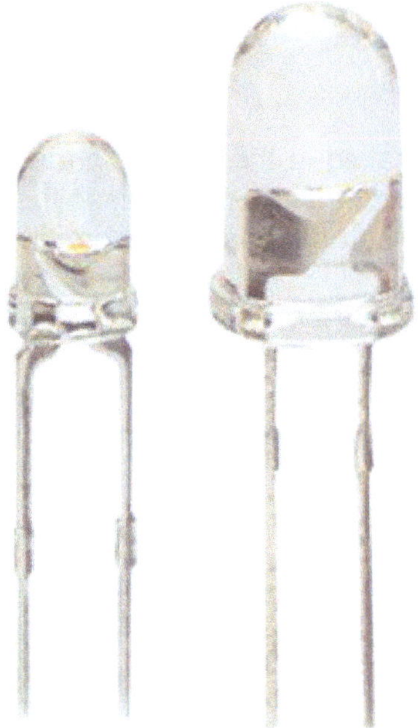

Figure 18: Typical white LED, which is used when you must produce an LED light in mountain areas.

Solar Collections

Now I want to describe another big project. Allan Aligo is his name, and he was not Catholic like most others, but more Protestant. Through connections in Australia, he developed projects.

He liked to drink beer and had been to several countries, including in Irish pubs, where they drank beer and talked about it. By chance I met him, and we did several projects together.

The most difficult and largest project was pumping water from a deep well. Via pipes on the ground, the water was pumped along a 2 km. long way up a hilltop.

As I recall, the top of the hill was approximately 30 to 50 meters high. A large tank was mounted on this hilltop.

Due to the height, the water pressure was enough to supply a very small village close to the hill. The locals had to provide free labor and provide certain materials.

Through churches and other donors, Allen raised money for projects from several places. He also knew many from Australia.

I can't remember where we bought the solar panels, but think they were German. The system was related to 4 large car batteries to a 48-volt system.

The hardest part was finding a suitable pump. Grundfos Manilla could not solve it. But we ended up getting hold of a 48-volt submersible pump, which was bought via Germany through an Asian supplier.

The German company was a competitor to Grundfos, but much better.

There was a British man called Kevin Irwin to whom I did a solar heating project.

He had built a house 10 km. from me, and he wanted access to hot water. We ended up making a solar collector made with copper pipes that was painted black.

There were ready-made plants that companies sold, but we made our own.

Most plants had a solar collector, where there was a tank above the solar collector. The tank was located on the roof. It was a system without pumps. So, the self-circulator - the ones that are used to this day all over the world.

Later it ended up with the tank being mounted inside the house and thus lower than the solar collector. This required a pump to move the energy down into the hot water tank.

In Manila, Grundfos had an importer. We needed a circulation pump, but the importer was impossible. We had to import it ourselves via Singapore and through a corrupt customs system.

Kevin then found out that he could buy it from England for less than half the price.

After much research and discussion, I ended up using my 12-volt Flojet pump and a transformer. A power supply was made as well as a small controller to start and stop the pump.

Since we were in a developing country in several ways, and Kevin wanted nice copper pipes, we couldn't find suitable fittings.

It ended up that we made our own pipe holders from waste from the pipes and painted them. This was the first time I made brackets and other difficult things from waste from the copper pipes we used and soldered with a gas torch.

Later he helped me with major problems in immigration. When my children and I were going to Denmark, we had to pay large sums in corruptions to the system.

Corruption was the biggest problem everywhere and those who don't know and have experienced this world don't understand the reality.

When the World Bank gave money to build new bridges, millions were given to the president, who then gives it to a governor, and the governor then gives money to a senator.

Every time this money changes hands, it disappears, and these politicians pocket up to 40 percent.

Since all bridges typically connect 2 municipalities, a municipality on each side of the bridge on the river must also receive corruption money for the mayor.

When I lived there, 60 percent of the GDP came to the country through relatives who sent money home, or from au-pair women and others who worked.

Therefore, the politicians do not want to offer the jobs locally, as there is more money for them in corruption when they send workers to countries far away. Typically, the governor gets the first 6 months' salary that the person earns, since he or she has obtained the connection.

This also happens today in many places in the world. If you are to offer aid to developing countries, I recommend that you don't just give a country money without reservation.

You must be there and ensure that the money does not end up with corrupt leaders and try to help spread democracy and, not least, use local raw materials if possible.

When I was there, I also tried to make my own solar panels. It was difficult and difficult to buy cells and expensive equipment for lamination.

Figure 19: My friend Ian Jordan's house with solar collectors.

I didn't have the capital, and many attempts failed, especially with encapsulation, where we used special silicone and epoxy.

Figure 20: A 12-volt panel of approx. 30 watts.

But in 2006, as you can see here in the picture where I am holding a panel in my hand, we succeeded in producing a 12-volt panel of approximately 30 watts.

Local materials from windows were used and I found a Russian company that delivered cells to me via DHL without customs corruption. It became the first functioning panel.

Special epoxy was used to encapsulate the cells, and with this we found a cheap method.

There was no one else at the time who produced their own panels in the country. Otherwise, it was the large foreign companies that dominated the market.

Later we managed to buy Q cells from Germany via another Asian country, and they had the world's best cells, and they were among the biggest producers in the world. But unfortunately, the company later went bankrupt.

I worked with a technical school close to where I lived. The school is called Tesda and can be compared to the Danish Mercantec in Viborg.

The school was in several places in the country, with its headquarters in Manila. The man with whom I worked well is called Francisco Jucar.

At the school, young people were trained in trades, and most of them, as previously described, got jobs abroad.

We tried to make solar drying ovens where you could dry crops and tobacco, of which there is a lot.

Since you need a large converter (inverter in English) of approximately 2000 watts to start the compressor on a standard refrigerator or freezer, it becomes difficult.

But I could remember that a company in Denmark near Viborg, Vibocold, as it is called now, had a possible solution.

At that time, the company was called Frigor, as I remember it. They made cool boxes for trucks that used the car's battery for cooling, just like a small refrigerator.

I called them from the Philippines and spoke to the person who is currently the export manager there.

It didn't work that time but can't remember why. What we needed was the Danfoss compressor, which can operate on both 12 and 24 volts from a car battery.

I have previously, immediately before I moved to the Philippines, delivered controls to Elcold in Hobro, Denmark, which makes deep freezers.

There was also an importer of Danfoss products in Manila at that time, but nothing would succeed with them in the same way as described with Grundfos.

But we managed to get Elcold in Hobro to send 3 pieces of 12-volt compressors from Danfoss via GLS to a place in Europe and from there on with Fedex to Manila and further on to our place.

They should be thanked for that as it worked miracles for us. The price for the compressor was high. Same price as a new fridge.

But when you replace the compressor and use the same thermostats as in the existing one, it works great.

It is almost silent and unlike a 220-volt normal compressor that uses 2000 watts for several seconds, when the compressor starts, this 12-volt compressor uses only 80 watts and after starting about 50 watts.

This eliminates the need for an expensive inverter, and energy consumption is low.

With a 50-watt solar panel and a 12-volt car battery it could run every day. We found a marine battery that has a long lifetime.

If you run the system on volts that fluctuate between 12 and 24 volts, the compressor can automatically switch between these.

We had 3 plants made, which ran well and reliably. Therefore, I think we must change behavior in all new construction and especially in public buildings and here uses this technique where possible.

Then again. The key word is small independent units, where possible and independence from large power grids and particularly useful in countries such as Africa, South America, and Asia.

If you combine it with a small wind turbine of perhaps 50 watts, the combination of sun and wind can provide independent energy and does not bother the neighbor due to its size.

I left the Philippines in August 2008 and my father had to pay for me and my 2 children's trip home as we had no funds. I was also shattered both physically and mentally, but this will come in another book later.

The future

I will now conclude and make my offer on what I think and believe will and must happen in the future. I have clairvoyant abilities and have had help from them many times and hope you understand that.

What I have described will not happen in a few years but it's necessary if we are to have a fairer world, and I am asking young people, especially the girls, to help here.

If we are to invent new techniques, we will also need biologists. An example is pumping water with almost no energy.

With pure energy, we should look at a tree or the blood vessels in our veins in our legs.

I imagine that we are imitating a large tree, which sucks up water from the ground and transports it out in the leaves. There are small flaps in a tree, which means that almost no energy must be used.

We must move away from pumps and instead imitate nature as in a tree. Therefore, biologists, which I think women are best at, must invent a new pumping system.

When the heart pumps blood around our body, and the blood must return from the legs up to the heart, there are small valves that open and close, so that absolutely minimal force is used.

Try to emulate these 2 things and in a few years, we will be using technology that resembles this.

In many years, I hope and believe that the weapons industry will no longer produce weapons as it does now but will produce aids that we need in renewable energy.

Let them produce robots that do all the grueling work and free human to socialize, abolish poverty and heal people.

In perhaps 50 years, I imagine that we will have developed a technique that resembles a solar cell but uses energy we do not know yet.

Right now, there must be light and not overcast to make energy with our well-known solar cells.

But if we use wavelengths that the naked eye can't see, the desk I'm sitting at will be able to provide all the energy we need, regardless of the weather and regardless of where we are on the globe.

Just think of UFOs, which can silently move quickly everywhere.

We live in the Stone Age seen with these future perspectives.

But until these prophecies become realities, we must do roughly as I have tried to explain. Hope you have enjoyed the book and have moved on to a better understanding of the future.

Sincerely, Jørn Nielsen November 2023

Epilogue

I have endeavored to be matter of fact and explain it in simple language. I have divided it into reasonably small and easy-to-read sections, so that you can take a short break along the way and not lose the overview.

During my retrospective, I have consulted several people – including some of my old friends, who have assisted me in refreshing my memory.

In this connection, I would like to thank Allen Aligo, and Francisco Jucar from the Philippines, as well as my old friend Ian Jordan, whom I had the great pleasure of visiting while I was writing the book.

I would also like to thank Britta from Tvind, whom I also visited and was sent informative knowledge and people mentioned in my book.

Next, I would like to say a big thank you to Elcold freezer in Hobro, which in its time helped me to get compressors from Danfoss, which were delivered by GLS and Fedex to the Philippines.

Also, thanks to Frigor, which today is called Vibocold, for elaborating on the story back in time when I was looking for a Danfoss compressor.

I expect the book to be available in technical schools and the country's libraries, so that young girls and boys will develop new forms of energy and thinking, so that I can help others to be ahead of their time.

If you only think in the present, no development will be possible, and society will not develop forward towards a human-friendly world seen globally.

Since I am spiritually inclined and use these qualities in my daily life, these abilities have also been used in the preparation of my book, which I hope will be a success.

In a future book, perhaps, I will work with the power of thought, as a possible energy.

Figure 22: The People's Center for Renewable Energy.

Figure 21: Tvind, Denmark Wind turbine.

Image overview